Assessment Standard for Sponge City Effects

A National Standard of the People's Republic of China

Assessment Standard for Sponge City Effects

GB/T 51345-2018

Written by Wenliang Wang, Linwei Zhang, Junqi Li, *et al.*

Translated by Linmei Nie, Haifeng Jia, Kefeng Zhang and Guangtao Fu

Published by IWA Publishing
Alliance House
12 Caxton Street
London SW1H 0QS, UK
Telephone: +44 (0)20 7654 5500
Fax: +44 (0)20 7654 5555
Email: publications@iwap.co.uk
Web: www.iwapublishing.com

First published 2020
© 2020 IWA Publishing

British Library Cataloguing in Publication Data
A CIP catalogue record for this book is available from the British Library

ISBN: 9781789060546 (paperback)
ISBN: 9781789060553 (eBook)

Foreword

Since its economic reform began 40 years ago, China's urbanization rate has continuously increased from less than 20% to about 60%, which is still accelerating. In many cities, the underground drainage infrastructure capacity is significantly lacking and does not match the fast-growing development of above-ground public infrastructure and buildings. The high imperviousness of the cities has significantly changed the urban water cycle and eco-systems, with consequences of frequent flooding, water pollution, ecosystem degradation, water shortage and other problems.

These urban problems are not unique to China. Developed countries try to adopt green and sustainable measures to solve these problems and to enhance urban water management in changing climate and urbanization, such as Green Infrastructure (GI), Sustainable urban Drainage Systems (SuDS) and Water Sensitive Urban Design (WSUD). China would learn from their experience, but rather to find a new path that is suitable for China's national condition. Therefore, in 2013, China initiated a national "**Sponge City Development**' Programme, as a novel concept of urban stormwater management and a new paradigm of urban development. China's Sponge City (SPC) builds on the successful experience on stormwater management from developed countries, but it also incorporates the traditional Chinese wisdom on water management, following the ancient philosophy of 'human-water harmony' and 'the highest goodness is like that of water'. In ancient China, people already had the early awareness to use rainwater resource. For example, the terraces were invented in the Qin Dynasty to make best use of rainwater. This concept was also introduced to the construction of towns and villages, such as 'Forbidden City' in Beijing and 'Fushougou' in Ganzhou, which is a unique drainage ditch for flood control, and the waterscape created by applying rainwater as the main water source in the classical Chinese gardens, which are all similar to the concept of modern urban stormwater management.

The SPC adheres to a problem-oriented principle and is designed systematically according to the concept of '**source reduction, process control and system remediation**.' According to local conditions, 'grey', 'blue' and 'green' measures should be used in a combined way. By integrating and adopting comprehensive measures such as '**infiltration, detention, retention, purification, utilization and discharge**', the natural ecosystems including the mountains, rivers, lakes, forests, farmland and grassland can be protected. Thus, this will bring multiple benefits, such as improving the quality of urban ecological environment, enlarging the capacity of urban flood prevention and mitigation, and enhancing urban amenity. As a result, the city can act like a 'sponge' which has high resilience in response to environmental changes and natural hazards.

In 2014, China launched two batches of national SPC development pilot projects in 30 cities. In the past 5 years, the pilot cities have systematically planned and promoted the SPC agenda according to the local conditions, and coordinated the implementation of sponge city with urban regeneration, flood control, treatment of malodorous water bodies, improvement of sewage treatment quality and efficiency, leading to a great amount of best engineering practices and guidelines, which could be valuable for other cities.

The review and assessment of the pilot effects and experience is extremely important for China to stimulate high quality SPC development in the next stage, enhancing the legislation, technical standardization, and promoting fundamental and theoretical research, professional training and new products. Therefore, at the end of 2018, the Ministry of Housing and Urban-Rural Development of the People's Republic of China (MHURD) released the '**Assessment Standard for Sponge City Effects**'. The publication of this standard is of great significance for China's continuous promotion of SPC development. It also disseminates Chinese knowledge and practices to the international community to exploitation of the innovative water management concepts to meet the challenges of global environmental change and urbanization.

The publication of the English version of the **Assessment Standard for Sponge City Effects** allows non-Chinese speaking readers to understand the Chinese assessment methods, which can be used as a reference by urban planners, designers, researchers, investment and management staff in development of modern urban stormwater systems.

This English edition is a translation of the original Chinese standard. During the translation and editing, we aimed to keep to the meaning of the original Chinese text. As such, this English edition is not intended to represent a different interpretation of the Chinese standard, but only to facilitate understanding of the Chinese standard for those who are not literate in Chinese language. Where this translation and editorial clarifications might differ from the original Chinese document, the reader is reminded that the Chinese document is the governing and only legally binding document.

During the process of drafting and translating the standard, we received strong support and constructive suggestions from the leading Chinese authors and officers of the Ministry of Housing and Urban-Rural Development (MHURD) and China Urban Water Supply and Drainage Association. We would like to thank MHURD for granting us the translation permission for this standard.

Thanks go to Mark Hammond and Niall Cunniffe, the Books Commissioning Editor of the IWA Publishing, London, UK for their efforts and suggestions.

Translation Authors:

Linmei Nie, Foundation CSDI WaterTech, Oslo, Norway.

Haifeng Jia, School of Environment, Tsinghua University, Beijing, China.

Kefeng Zhang, Water Research Centre, School of Civil and Environmental Engineering, the University of New South Wales, Sydney, Australia.

Guangtao Fu, The Centre for Water Systems, University of Exeter, UK.

Reviewers:

Wenliang Wang, Junqi Li, Leading authors of the Chinese version of this assessment standard, Beijing University of Civil Engineering and Architecture, Beijing, China.

Linwei Zhang, Leading author of the Chinese version of this assessment standard, former General Director of the Construction Department, the Ministry of Housing and Urban-Rural Development, now China Urban Water Association, Beijing, China

Linus Zhang, Dept. Water Resources Engineering, Lund University, Sweden
Alberto Campisano, Department of Civil Engineering and Architecture, the University of Catania, Italy

Ben Urbonas, Urban Watersheds Research Institute, Denver, Colorado, USA

James C.Y. Guo, Urban Watersheds Research Institute, Denver, Colorado, USA

Bob Brashear, Center for Infrastructure Modeling and Management, Austin, Texas USA

UDC

A National Standard of the People's Republic of China

P   **GB/T 51345-2018**

Assessment Standard for Sponge City Effects

Published on 26-12-2018   adopted from 01-08-2019

Jointly published by
The Ministry of Housing and Urban-Rural Development (MHURD), P.R. China
The General Administration of Quality Supervision, Inspection and Quarantine, P. R. China

Preface

According to the requirement of the *Letter on drafting 'Assessment Standard for Sponge City Effects'* by the Ministry of Housing and Urban – Rural Development (MHURD [2016] No. 12) the development group produced this standard which was established on the basis of extensive investigation, practical experience, the relevant international standards and extensive consultation.

The main technical contents of this standard include 1) general provisions; 2) terms and symbols; 3) basic requirements; 4) assessment items and 5) assessment methods.

This standard is administratively managed by the Ministry of Housing and Urban – Rural Development. China Construction Science and Technology Group Co., Ltd. is responsible for the interpretation of the technical contents. During implementing the standard, any comments or suggestions should be directed to the China Construction Science and Technology Group Co. Ltd. at the following address:

China Construction Science and Technology Group Co., Ltd.
Block A, Desheng Kaixuan Building
Deshengmenwai Avenue No. 36
Xicheng District, Beijing 100044
P.R. China.

The leading organizations of the standard editing group are:

- China Construction Science and Technology Group Co., Ltd
- China Urban Water Association
- Beijing University of Civil Engineering and Architecture

With the participation of the following organizations:

- China Academy of Urban Planning and Design
- Urban Water Management Office, the Ministry of Housing and Urban – Rural Development
- Institute of Standard Quota, the Ministry of Housing and Urban – Rural Development
- Urban Planning and Design Institute Co., Ltd. of Shenzhen
- Beijing General Municipal Engineering Design and Research Institute Co., Ltd.
- Shanghai Municipal Engineering Design Institute (Group) Co., Ltd.
- North China Municipal Engineering Design and Research Institute Co., Ltd.
- China Urban Construction Design and Research Institute Co., Ltd.
- Science and Technology Development Promotion Center, the Ministry of Housing and Urban – Rural Development
- Zhejiang Province Institute of Architectural Design and Research

The contributory authors are:

Wenliang Wang, Linwei Zhang, Junqi Li, Wei Chen, Huiwei Xu, Chengjiang Li, Yong Chen, Yufen Shu, He Wen, Weilan Bai, Wu Che, Zheng Yang, Yongwei Gong, Wei Zhang, Xinxin Ren, Hongtao Ma, Xiyan Ren, Jiazhuo Wang, Li Zhao, Dongquan Zhao, Lu Yu, Wei Gao, Yingjun Hu, Ye Zhao, Yunfeng Shen, Yang Zhao, Jianlong Wang, Sisi Wang, Kun Mao, Xiaoli Du, Xuwei Liu, Guoyu Wang, Kuang Sheng, Yongpeng Lv, Yan Chen, Xiangjuan Kong, Yong Liang, Jiang Cheng

The reviewers are:

Nanqi Ren, Li'an Hou, Jun Xia, Kebai Zheng, Jun Sui, Xiang Liu, Xiaoxin Zhang, Jiahong Liu, Hailong Liu and Qingyuan Tong

Contents

1 GENERAL PROVISIONS

1.0.1 Sponge City (SPC) is an important measure to implement the concept of ecological and green development in urban areas of China. It helps to promote the systemic development and construction of urban infrastructure recognizing cities as a living community, where human and nature live in harmony. This standard is thereby developed in order to promote the development of Sponge City, improve the quality of urban ecological environment, enhance the capability of urban disaster prevention and mitigation, increase the supply of high-quality ecological products and enhance people's satisfaction and happiness, and manage the assessment of the Sponge City infrastructure effects.

1.0.2 This standard is applicable to the assessment of all Sponge City effects.

1.0.3 The assessment of Sponge City infrastructure should comply with the purpose of Sponge City infrastructure requirements, aiming to:

- conserve the natural ecosystems of mountains, rivers, lakes, forests, farmland and grassland,
- preserve the hydrological characteristics that are fundamental properties of the ecosystems, including infiltration, retention, evaporation (evapotranspiration) and runoff, and
- protect and restore the capacity of natural storage, infiltration and purification of rainfall runoff process.

1.0.4 The assessment of Sponge City infrastructure should comply with the technical approach and method of Sponge City development based on the defined development goals which are, in turn, based on specific project site conditions. Further, assessments should be planned systematically according to the philosophy of 'source reduction, process control and systematic remediation'. Starting from the project site conditions, it should assess any combination of grey and green infrastructure which adopt integrated measures of *'infiltration, detention, retention, purification, utilization and discharge'*.

1.0.5 In addition, to meet the requirements of this standard, the assessment of Sponge City effects should comply with the relevant regulations of any currently prevailing national standards.

2 TERMS AND SYMBOLS

2.1 Terms

2.1.1 Sponge city (SPC)

Through a combination of urban planning and constructability, the Sponge City perspective on urban drainage is developed based on a systematic approach of *source reduction, process control, and systematic remediation*, and it adopts comprehensive technical measures of *infiltration, detention, retention, purification, utilization and discharge* of stormwater. Sponge City infrastructure coordinates systematically the different aspects of water quantity and quality, ecology and safety. It aims to achieve the multiple objectives of mitigation of urban flooding, control of runoff pollution, improvement of urban water environment and restoration of urban water ecology, thus it provides solid foundation for the systematic management of mountains, water systems, forests, lakes and lands, promotion of green development and development of a beautiful China.

2.1.2 Volume capture ratio of annual rainfall

The ratio of the average annual rainfall volume controlled through natural and artificial infiltration, retention, purification and other measures, as compared to the average total annual rainfall volume.

2.1.3 Sponge effect

The effects of conservation and restoration of natural hydrological characteristics achieved by Sponge City infrastructure.

2.1.4 Catchment

The area contributing to runoff collection delineated according to the topography or determined by the drainage channels and pipe networks.

2.1.5 Overflow outlet

The facility through which urban runoff is naturally discharged after the volume capacity of sponge city infrastructure is exceeded.

2.1.6 Green Infrastructure (GI)

The urban runoff control facilities with natural or artificial ecological systems.

2.1.7 Grey infrastructure

Traditionally engineered drainage facilities with high energy consumption levels.

2.1.8 Impervious land surface ratio

The ratio of the impervious land surface area, excluding the roof area, compared to the total land surface area.

2.1.9 Urban water body

The natural and man-made water bodies such as rivers, lakes, wetlands and ponds in the urban planning area.

2.2 Symbols

2.2.1 Calculation of the facility's runoff volume control and volume capture ratio of annual rainfall

A — effective infiltration area.

J — hydraulic gradient.

K — the saturated hydraulic conductivity of the soil or artificial media.

T_d — the designed emptying time of the facility.

t_p — the discharge duration during a rainfall event.

t_s — the infiltration duration during a rainfall event.

V_{ed} — the runoff control volume controlled by the extended detention facility.

V_{in} — the runoff volume controlled by the infiltration, filtration and retention facilities.

V_s — the effective retention volume of the facility.

W_{ed} — the discharge volume from extended detention facility during a rainfall event.

W_{in} — the infiltrated volume by infiltration, filtration and retention facilities during a rainfall event.

α — volume capture ratio of annual rainfall.

φ — runoff coefficient.

2.2.2 Calculation of the variation trend of groundwater depth

Δh_1 — the average drop of groundwater level in the urban built-up area before construction of Sponge City infrastructure.

Δh_2 — the average drop of groundwater level in the urban built-up area after construction of the Sponge City infrastructure.

2.2.3 Calculation of the urban heat island effect mitigation

ΔT_1 — the average daily temperature difference in urban areas and the surrounding suburbs before the construction of Sponge City infrastructure.

ΔT_2 — is the average daily temperature difference between the urban areas and the surrounding suburbs after the construction of Sponge City infrastructure.

3 BASIC REQUIREMENTS

3.0.1 The assessment of Sponge City effects should be based on the land use of the urban built-up area, to evaluate the integrated 'sponge' effects of the source reduction projects, drainage catchments, and the whole built-up areas.

3.0.2 The assessment of sponge city effects should use the drainage catchment as a basic statistical unit to calculate the proportion of the urban built-up area that meets the requirements of this standard to the total area of the urban built-up area.

3.0.3 The assessment of Sponge city effects is divided into **Required** and **Optional**. The urban built-up area should satisfy all **Required** effects of this standard. The **Optional** effects should be evaluated, but their evaluation results do not affect the assessment conclusions.

3.0.4 The assessment of Sponge City effects should be conducted comprehensively combining on-site inspection, project document review and model simulations, based on continuous monitoring data of no less than one year on the typical projects, pipe not works and urban water bodies.

3.0.5 The assessment of the implementation effectiveness of the associated Sponge City source reduction projects should be based on the development objectives and by monitoring and evaluation of representative projects. For each type of the representative projects, one or two monitoring projects should be selected to monitor the water volume and water quality at overflow outlets or inspection wells connecting to the municipal pipe network or water bodies.

4 ASSESSMENT ITEMS

4.0.1 The assessment of the effects of Sponge City development should evaluate the benefits of projects, and whether the sponge effects have been achieved. Table 4.0.1 provides an overview of provisions of the assessment items and requirements.

4.0.2 Of the assessment items of Sponge City effects, **V**olume capture ratio of annual rainfall and runoff volume control, **I**mplementation effectiveness of the source reduction project, **C**ontrol of road surface ponding and local flood, **U**rban water quality and **N**atural ecological pattern control and shoreline for ecological conservation are **R**equired indicators for assessment, while the **V**ariation trend of groundwater depth and the effect mitigation of **U**rban heat island effect are **O**ptional indicators for assessment.

Table 4.0.1 Assessment items and requirements for sponge city infrastructure

Assessment item		Assessment requirement	Assessment method
1. Volume capture ratio of annual rainfall and runoff volume control		(1) For new development areas, the volume capture ratio must be not smaller than the lower limit of the value specified according to the 'Zoning map of volume capture ratio of annual rainfall in China' (Figure 4.0.1), and runoff control volume must be not smaller than the corresponding runoff volume calculated using the volume capture ratio. (2) For re-development areas, according to technical and economic comparison, the capture ratio is not recommended smaller than the lower limit specified according to the 'Zoning map of the volume capture ratio of annual rainfall in China' (Figure 4.0.1), and the corresponding runoff volume calculated from the volume capture ratio.	Should meet the requirements set out in this standard 5.1
2. Implementation effectiveness of the source reduction project	Residential area	(1) Volume capture ratio of annual rainfall and the runoff volume control: (i) for new development projects, the ratio should be not lower than the lower limit of the value specified according to the 'Zoning map of the volume capture ratio of annual rainfall in China' (Figure 4.0.1), and the calculated corresponding runoff volume; and (ii) for redevelopment projects, according to technical and economic comparison, the ratio is not recommended lower than the lower limit of the value specified according to the 'Zoning map' (Figure 4.0.1), and the calculated corresponding runoff volume, or meet the similar control requirements of urban planning. (2) Runoff pollution control: annual runoff pollutant reduction (in terms of Suspend Solids - SS) is not recommended less than 70 percent for new development projects and 40 percent for redevelopment projects or meet the similar control requirements of urban planning. (3) Peak runoff control: for the design storm events of minor and major drainage systems, the peak runoff discharge is not recommended exceed the original value before the development of the new development projects; for redevelopment projects, the peak runoff discharge must not exceed the value before the redevelopment. (4) Impervious land surface ratio is not recommended larger than 40 percent for new development projects; and should not be larger than the original ratio for redevelopment projects, but not recommended larger than 70 percent.	Should meet the requirements set out in this standard 5.2
	Roads, parking lots and open squares	(1) Roads: runoff pollution should be controlled according to the requirements of planning and design of the projects. For roads with flood prevention and drainage functions, the drainage function and capacity should be maintained.	

(Continued)

Table 4.0.1 Assessment items and requirements for sponge city infrastructure *(Continued)*.

Assessment item	Assessment requirement	Assessment method

	(2) Parking lots and open squares:
	  (i) Volume capture ratio of annual rainfall and runoff volume control:
	    (a) for new development projects, the ratio should be not smaller than the lower limit of the value specified in the 'Zoning map' (Figure 4.0.1), and the corresponding runoff volume calculated using the volume capture ratio; and
	    (b) for redevelopment projects, the ratio is not recommended lower than the lower limit of the value specified in the 'Zoning map' (Figure 4.0.1), and the corresponding runoff volume calculated.
	  (ii) Runoff pollution control: annual runoff pollutant reduction (in terms of SS) is not recommended less than 70 percent for new development projects, and 40 percent for redevelopment projects.
	  (iii) Peak runoff control:
	    (a) for the design storm events for minor and major drainage systems, the peak runoff discharge is not recommended exceed the original value before the development of the new development projects;
	    (b) for redevelopment projects, the peak runoff discharge must not exceed the value before the redevelopment.	

Park and protective green space	(1) For new development projects: the runoff control volume must be not smaller than the calculated runoff volume corresponding to the volume capture ratio of annual rainfall equals to 90 percent; and for redevelopment projects, after economic and technical comparison, the runoff volume control is not recommended less than the calculated runoff volume subject to the volume capture ratio of annual rainfall equals to 90 percent. (2) The area should receive the runoff from surrounding areas according to the requirements of planning and design of the projects.	
3. Road surface ponding and local flood control	(1) Combine grey and green infrastructure solutions in a way that maximizes the role of green infrastructure facilities in detention and reduction of the peak runoff values, and (2) Surface ponding should not occur under the design storm events of minor drainage system. (3) There must be no local flooding under the design storm events of major drainage system.	Should meet the requirements set out in this standard 5.3

(Continued)

Table 4.0.1 Assessment items and requirements for sponge city infrastructure (*Continued*).

Assessment item	Assessment requirement	Assessment method
4. Urban water quality	(1) The grey and green infrastructure solutions should be combined reasonably to fully utilize the role of green infrastructure in the control of runoff pollution, Combined Sewer Overflows (CSOs) and water quality purification. (2) During dry weather conditions, no sewage or wastewater is discharged directly without treatment. (3) During wet weather conditions, the pollution discharges from separated sewer systems because of mis-connections or CSOs must be controlled, and should not lead to malodorous water bodies; or the discharges from overflow outlets of the mis-connected separate systems and CSOs should be controlled, such that the volume control rate of annual overflow should be not less than 50 percent, and the monthly average SS concentration discharged from overflow treatment facilities should not exceed 50 mg/L. (4) No malodorous waterbody: transparency >25 cm (in case the water depth is less than 25 cm, the index value is taken as 40 percent of the water depth), dissolved oxygen >2.0 mg/L, redox potential >50 mV, ammonia nitrogen <8.0 mg/L. (5) Should be not inferior to the water quality before Sponge City infrastructure is installed. When there is upstream inflow to the river system, the water quality of the downstream river cross sections is not recommended worse than the upstream inflow water quality during the dry weather conditions.	Should meet the requirements set out in this standard 5.4
5. Natural ecological pattern management and shoreline for ecology conservation	(1) The surface area of natural waterbodies is not recommended reduced before and after urban development, and the natural topography and landscape pattern should be restored and rehabilitated maximally. Natural flood channels, flood plains and ecologically sensitive areas, such as wetlands, forest and grass lands must not be occupied; or the management and control requirements of the blue green lines should be met, according to urban planning. (2) The ecological shoreline of the new and redevelopment of urban water bodies is not recommended not less than 70 percent, except for the dock for production and necessary shorelines reserved for flood control and protection according to urban planning.	Should meet the requirements set out in this standard 5.5
6. Variation trend of the groundwater depth	The descending trend of annual average groundwater phreatic level should be alleviated.	Should meet the requirements set out in this standard 5.6
7. Urban heat island effect mitigation	The average daily temperature difference between the urban areas and the surrounding suburbs during summer seasons (June to September) should have a descending trend, comparing to the same period in history (subtracting the impacts of natural temperature variation).	Should meet the requirements set out in this standard 5.7

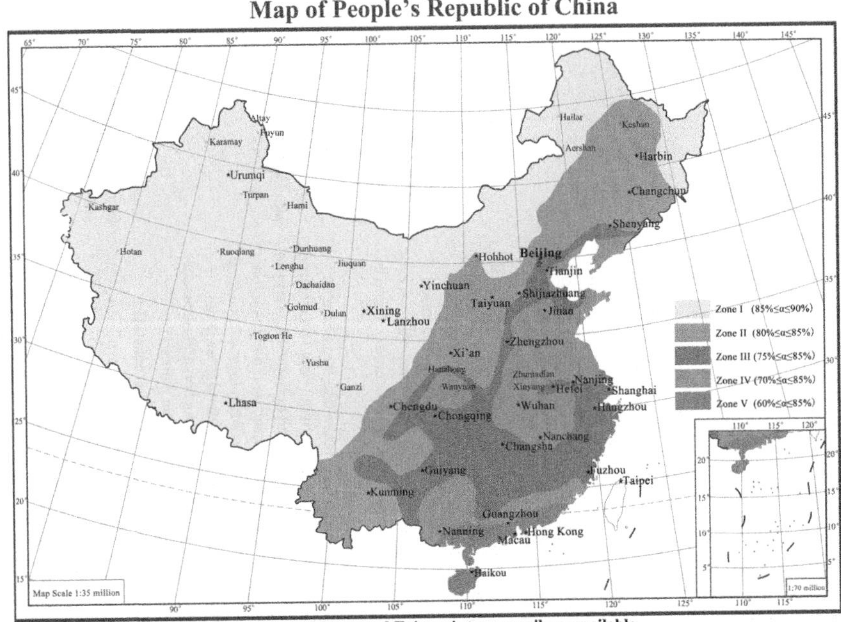

Note:the site information of Hong Kong, Macau and Taiwan is temporarily unavailable.

Figure 4.0.1 Zoning map of volume capture ratio of annual rainfall in China.

5 ASSESSMENT METHODS

5.1 Volume capture ratio of annual rainfall and runoff volume control

5.1.1 The volume capture ratio of annual rainfall and the runoff control volume should be assessed through auditing of the facility's runoff control volume, monitoring, model simulation and on-site inspection.

5.1.2 Estimation of the facility's runoff control volume should meet the following requirements:

(1)   According to the design rainfall depth, which is related to the volume capture ratio of annual rainfall, and the facility's catchment area, a specific volume-based method is used to calculate the runoff control volume for infiltration and retention facilities. A separate method is used for extended detention facilities. On-site inspection should be carried out to verify that the actual runoff control volume of each facility meets the design requirements.

(2)   For infiltration, filtration and retention facilities, the runoff control volume should be calculated according to the following equations:

$$V_{in} = V_s + W_{in} \qquad\qquad (5.1.2\text{-}1)$$

$$W_{in} = KJAt_s \qquad\qquad (5.1.2\text{-}2)$$

where:

V_{in} — the runoff volume controlled by the infiltration, filtration and retention facilities, m^3.

V_s — the effective retention volume of the facilities, m^3.

W_{in} — the infiltrated volume by infiltration, filtration and retention facilities during a rainfall event, m^3.

K — the saturated hydraulic conductivity of the soil or artificial media, m/h. It is calculated according to the effective retention depth of the retention zone and design emptying time of the facility. It depends on the soil type or artificial media compositions. The hydraulic conductivity parameters of different types of soils can be determined according to the provisions in the national standard of 'Technical code for rainwater management and utilization of building and sub-district' (GB50400)[1].

J — hydraulic gradient. A default value is set to 1.

A — effective infiltration area, m^2.

t_s — the infiltration duration during a rainfall event, h. It is the local average rainfall events duration. If the data is not available, a default value between 2 h and 12 h can be used according to the characteristics of the average rainfall events duration.

(3)　For the extended detention facility, the runoff control volume is calculated according to the following equations:

$$V_{ed} = V_s + W_{ed} \tag{5.1.2-3}$$

$$W_{ed} = (V_s/T_d)t_p \tag{5.1.2-4}$$

where:

V_{ed} – the runoff volume controlled by the extended detention facility, m^3.

W_{ed} – the discharge volume from the extended detention facility during a rainfall event, m^3.

T_d – the designed emptying time of the facility, h. It is calculated according to the required time for settlement of suspended solids (SS).

t_p – the discharge duration during a rainfall event, h. It is the local average rainfall events duration. If the data is not available, a default value between 2 h and 12 h can be used according to the characteristics of the average rainfall events duration.

5.1.3 The determination of the appropriate volume capture ratio of annual rainfall of the project should meet the following requirements:

(1)　For each facility, on-site inspection should be carried out to quantify the actual runoff control volume, which is then used to estimate the corresponding rainfall control depth. Thereafter the actual volume capture ratio of annual rainfall can be estimated with reference to 'Schematic diagram of the relationship between the volume capture ratio of annual rainfall α and the design rainfall depth H' (Figure 2 of the Article description).

(2)　The appropriate volume capture ratio of annual rainfall of the whole project should be determined by calculating the weighted average values of all the facilities and underlying surface that are not controlled by any facility with their service catchment areas (including their own areas) and the underlying surface areas as weights.

(3)　For the impervious underlying surface that is not controlled by any facility, the volume capture ratio of annual rainfall is zero;

(4)　For the pervious underlying surface that is not controlled by any facility, the volume capture ratio of annual rainfall shall be determined by either referring to 'Schematic diagram of the relationship between the volume capture ratio of annual rainfall α and the design rainfall depth H' with the design rainfall depth being the depth of incipient and subsequent loss (i.e. sum of losses from plant interception, incipient field storage and infiltration), or simply applying the following equation:

$$\alpha = (1 - \varphi) \times 100\% \tag{5.1.3}$$

where:

α – volume capture ratio of annual rainfall, in percentage.

φ – runoff coefficient.

5.1.4 For the projects monitored, the volume capture ratio of annual rainfall can be assessed using the following methods:

(1) On-site inspection should be conducted to investigate the actual overflow discharges of the facility that meets the design requirements of runoff volume control through the measures of '*infiltration, detention, retention, purification and utilization*'.

(2) 'Time-flow' data should be monitored at the overflow outlet or an inspection manhole that connects to the municipal sewer network and collected through automatic and continuous monitoring over at least one rainy season or one flood season.

(3) At least two rainfall events should be selected with rainfall depths approximately ± 10 percent of the design rainfall depth and have the antecedent dry periods longer than the emptying time of the facility. For these events, if there is no discharge flow at the overflow outlet or the inspection manhole that are connected to the municipal sewer network, or the discharge flow equals to the flow from the facility through purification via filtration and sedimentation, the project can be evaluated as meeting the design requirements.

5.1.5 For the catchment, the volume capture ratio of annual rainfall should be assessed according to the following requirements:

(1) Simulation models should be used to estimate the volume capture ratio of annual rainfall.

(2) The model should have the capability to simulate rainfall-runoff, pipe flows and source reduction facilities, etc.

(3) For model simulation, the following information should be collected: parameters of the source reduction facilities, sewer network topology and pipe defects, catchment characteristics, terrain, as well as a minimum of the latest 10-year continuous rainfall time series with time resolutions of 1 min or 5 min or 1 h.

(4) To calibrate and validate the models, together with the projects monitored within the catchment, at least one year continuous flow time series at the municipal drainage outlet or observed water level time series at the inlet of the pumping station should be collected for at least one typical catchment, with flow meters installed at the end of the municipal drainage outlet and key nodes of the network upstream. Model calibration and validation should be done by selecting monitoring data of at least two events respectively with the maximum 1 h rainfall depth equivalent to the design rainfall depth of the minor system. Nash-Sutcliffe coefficients for model calibration and validation should be no less than 0.5.

5.1.6 For urban build-up areas (areas of a city which have already developed), the volume capture ratio of annual rainfall should be determined by calculating the weighted average value of the individual catchments with their areas as weights.

5.2 Implementation effectiveness of the source reduction project

5.2.1 For the projects in residential areas, the implementation effectiveness should be assessed according to the following requirements:

(1) The volume capture ratio of annual rainfall and its corresponding runoff volume control should be assessed according to the method and requirements specified in section 5.1;

(2) To assess the effect of runoff pollution reduction, an integrated approach of reviewing design and construction documents together with on-site inspection should be applied. The design structure of the facility, runoff control volume, emptying time, operation condition, plant types are checked thoroughly to control whether the SS reduction meets the requirements, provided that the emptying time should be no longer than the wet tolerance duration of the plants. For solids and oil removal from a specific facility, the water treatment capability should meet the design requirements. For the new development project, it is recommended to have runoff pollution control facilities installed for all the underlying impervious surfaces; while for the redevelopment projects, the ratio of the controlled impervious surface area to the total area of the impervious surface is better no less than 60 percent.

(3) The peak runoff value control effect should be assessed according to the model simulation, design and construction documents review and on-site inspection.

(4) The impervious land surface ratio should be assessed by reviewing design and construction documents, in combination with onsite inspection.

5.2.2 For the project with regards to roads, parking lots and open squares, the implementation effectiveness should be assessed according to the following requirements:

(1) The volume capture ratio of annual rainfall and the corresponding runoff volume control should be assessed according to the methods specified in section 5.1;

(2) The runoff pollution and peak runoff value control should be assessed according to the methods specified in 5.2.1;

(3) To assess the road drainage and flood prevention performance, a combined approach of reviewing design and construction documents and on-site inspection should be applied.

5.2.3 For the project with regards to urban parks and protective green space, the implementation effectiveness should be assessed according to the following requirements:

(1) The volume capture ratio of annual rainfall and the corresponding runoff volume control should be assessed according to the methods specified in section 5.1;

(2) The control effect of urban rainfall-runoff through the urban parks and protective green spaces, an integrated approach of reviewing design and construction documents with on-site inspection should be used in order to check whether the control catchment area and the implementation scale of the facilities meet the requirements.

5.3 Control of road surface ponding and local flooding

5.3.1 To assess the suitable combination of grey and green infrastructure, a combined approach of reviewing the design and construction documents and on-site inspection should be used.

5.3.2 To assess the control of road surface ponding, the approaches of reviewing design and construction documents, and video recordings should be used according to the following requirements:

(1) Reviewing the design and construction documents, the design runoff depth should not be higher than 15 cm for the road ditches and the lower points of the key flood-prone sites.

(2) Select the design storm/rainfall events by screening the storm event with the maximum 1-hour rainfall depth of no lower than the design rainfall depth of the minor system specified in the national standard *"Code for design of outdoor wastewater engineering" (GB 50014)*[2], and then analyze the video recording to ensure that the runoff depth is not higher than 15 cm for the road ditches and the lower points of the key flood-prone sites, and the water retreating time should not be longer than 30 min.

5.3.3 For the assessment of local flood control, a combined approach of reviewing video recordings, on-site inspection and model simulation should be applied, and comply with the following requirements:

(1) The models should have the capability to simulate rainfall-runoff, pipe flows, overland flow, rivers, lakes and other natural waterways.

(2) To set up the model for simulation, the following information should be collected: sewer network topology and pipe defects, catchment characteristics, terrain and ponding monitoring data of the key flood-prone sites, as well as the data of design rainfall distribution for major drainage system, with a minimum temporal resolution of 5 min and total duration of 1440 min.

(3) To calibrate and validate the model parameters, selecting at least one typical catchment with video monitoring instruments installed at flood-prone points, and flow meters installed at the municipal drainage outlet and key

nodes of the network upstream, monitoring over at least one year of the ponding area, depth and the water retreating time at the flood-prone points based on the analysis of video recording, the continuous flow time-series data at the municipal drainage outlet or water level time-series data at the inlet of pumping station. The model should be calibrated and validated by selecting monitoring data of at least two events respectively with maximum 1-hour rainfall depth of no less than the design rainfall depth of the minor drainage system. Nash-Sutcliffe coefficients for model calibration and validation should be not less than 0.5.

(4)   Using the design storm of the major system to simulate the road surface ponding area, ponding depth and water retreating time, the values of which should all meet the current national standard of 'Technical code for design of outdoor wastewater drainage system' (GB 50014)[2] and 'Technical code for urban flooding prevention and control' (GB51222)[3].

(5)   Review the collected video recording data of no less than 1 year, and for the actual rainfall events that have a maximum 1-hour rainfall depth of no less than the design rainfall depth of the major drainage system. The road surface ponding area, ponding depth and water retreating time should be analyzed, and meet the requirements of the current national standards, i.e., 'Technical code for design of outdoor wastewater drainage system' (GB 50014)[2] and 'Technical code for urban flooding prevention and control' (GB51222)[3].

5.4 Urban water quality

5.4.1 For the assessment of the connection of grey and green infrastructure, a combined approach of reviewing the design and construction documents along with on-site inspection should be used.

5.4.2 For the assessment of the control of sewage and wastewater discharged directly without treatment during dry weather, on-site inspection should be carried out to ensure that no sewage and wastewater are discharged directly from outlets of municipal drainage network during dry weather.

5.4.3 For the assessment of the mitigating overflows of separated sewer systems or overflows from combined sewer systems, a combined approach of reviewing design and construction documents, monitoring, model simulation as well as on-site inspection should be used and should also comply with the following requirements:

(1)   Review the project design and construction documents and make on-site inspections to investigate the proper implementation of overflow control measures.

(2)   Monitor the concentration of SS discharge from overflow treatment facilities, with at least one sample taken each discharge event.

(3) The reduction in the volume of the annual overflows shall be assessed by model simulation or monitoring approach. The assessment model should have the functionality to simulate rainfall-runoff, pipe flows and source reduction facilities, etc. To set up the model for simulation, the following information should be collected: parameters for the source reduction facilities, sewer network topology, including hydraulic impacts of pipe defects, operational conditions of the intercepting main sewer pipe and sewer treatment plant, catchment characteristics, terrain and continuous rainfall data of the most recent 10 years with time-steps of 1-minute, 5-minute or 1-hour. When performing an assessment based on monitoring data, flow time series data of at least the most recent 10 years at the outlets of overflow should be used for the assessment.

(4) The calibration and validation of the model parameters should follow the provisions described in section 5.1.5 of this standard. At least two rainfall events with maximum 1-hour rainfall depth that are almost equal to the design rainfall depth of the minor drainage system should be selected, and the time-flow monitoring data at CSO outlets are to be used for calibration and validation. The overflow volume at the CSO outlets over at least 10 years should be simulated through the model or calculated according to the monitoring data before and after the implementation of measures including correction of the misconnection of the stormwater and wastewater pipes, interception, detention, retention and treatment.

5.4.4 The assessment of malodorous waterbodies and monitoring of water quality should meet the following requirements:

(1) The testing methods for water quality indicators should comply with the provisions of the current national standard 'Examination methods for municipal sewage' (CJ/T51)[4].

(2) Monitoring points should be set every 200–600 meters along the waterbody; for the rivers that receive upstream water flows, monitoring points should be set in correspondence of both upstream and downstream cross-sections. At least three points should be monitored of each water body. The sampling point should be set at 0.5 meters below the surface. In case that the total water depth is lower than 0.5 meters, the sampling points should be set at one-half of the water depth.

(3) Sampling should be conducted at least once in one to two weeks' period, and at least once within one day after a medium-size rainfall event, and it should continue for one year. Alternatively, continuous monitoring for at least 40 days in both wet and dry seasons should be done and one sample per day should be taken.

(4) The monthly average values of water quality parameters at each of the monitoring points should meet the requirements given in Table 4.0.1 of this standard.

5.5 Natural ecological pattern management and shoreline for ecology conservation

5.5.1 For assessment of natural ecological pattern management, documents providing detail on natural ecological patterns as well as on-site inspection should be conducted, complying with the following requirements:

(1) Review any urban master planning, relevant specific planning and regulation documents for the urban blue and green line protection, as well as the high-resolution remote sensing image before the urban development and at the current situation.

(2) Inspect on-site the ecological sensitive areas, including nature landscape pattern, nature flooding channels, floodplains, wetlands, forest and grasslands, as well as the blue green line and ecological red line management areas.

(3) After the watershed has been developed, surface area of natural water bodies should not be reduced, the natural topography and landscape pattern should not be changed, and the ecological sensitive areas, such as nature flood channels, floodplains and wetlands should not be invaded; or the relevant planning requirements are met.

5.5.2 Assessment of ecology conservation shoreline should be conducted by reviewing the design and construction documents for the new or redevelopment urban water projects, and specifying clearly the length and ratio of the shoreline for ecology conservation. The implementation conditions should be inspected on site.

5.6 Variation trend of groundwater depth

5.6.1 The variation of groundwater levels for unconfined aquifers should be monitored in the urban build-up area over the most recent five years before Sponge City infrastructure, and at least one year of monitoring after Sponge City infrastructure has been implemented.

5.6.2 The monitoring of groundwater levels should follow the provisions in the current national standard 'Technical code for groundwater monitoring' (GB/T 51040)[5].

5.6.3 Compare Δh_1 and Δh_2 and ensure $\Delta h_1 - \Delta h_2 > 0$, where Δh_1 is the average drop of the groundwater level in the urban built-up area before Sponge City infrastructure, and Δh_2 is the average drop of the groundwater level in the urban built-up area after the Sponge City infrastructure. In other words, the groundwater level should rise after the Sponge City infrastructure.

5.6.4 If there is only one year of ground water level monitoring after Sponge City infrastructure, compare Δh_1 and Δh_3 and ensure $\Delta h_1 - \Delta h_3 > 0$, where Δh_3 is the average drop of groundwater level in the year of Sponge City infrastructure comparing to the previous year. Alternatively, it is acceptable if the groundwater level rises after the Sponge City infrastructure.

5.7 Urban heat island effect mitigation

5.7.1 The temperature changes in the urban areas and their surroundings suburbs should be monitored. Temperature monitoring methods should comply with the provisions in the current national standard 'Specifications for surface meteorological observation – Air temperature and humidity' (GB/T 35226)[6].

5.7.2 The temperature monitoring data before the Sponge City infrastructure should cover the average daily temperatures from June to September for at least the recent five years. The temperature monitoring data after the Sponge City infrastructure should be the average daily temperatures from June to September for at least one year.

5.7.3 Compare ΔT_1 and ΔT_2 to ensure ΔT_2 is smaller than ΔT_1, where ΔT_1 is the average daily temperature difference between the urban areas and the suburbs before Sponge City infrastructure; ΔT_2 is the average daily temperature difference between the urban areas and the suburbs after Sponge City infrastructure.

DESCRIPTION OF WORDS AND EXPRESSIONS IN THIS STANDARD

(1) In order to distinguish the exact meaning of the requirements in the implementation of the provisions of this standard, the words and expressions apply in this standard:

   (i) When it is very strict, and must do: 'Must' is used for positive actions, and 'be strictly prohibited' is used for opposite behaviors.

   (ii) When it is strictly enforced, something should be done under normal circumstances: 'Should' is used for positive actions, while 'should not' is used for the opposite behaviors.

   (iii) Provided with some flexibility, one should do so first when the conditions are fulfilled. 'is recommended' is used for positive actions, while 'is not recommended' is used for any negative behaviors.

   (iv) When having some choices, and it is allowed under certain conditions, 'may' is used.

(2) When it is indicated that it should comply with the requirements specified in other relative standards and codes, the expressions of 'it should meet the requirements of …', or 'comply with the requirements of …'.

LIST OF CITED STANDARDS

[1] MHURD and AQSIQ, 2016. Technical code for rainwater management and utilization of building and sub-district (GB50400). The Ministry of Housing and Urban-Rural Development (MHURD); the General Administration of Quality Supervision, Inspection and Quarantine (AQSIQ), P. R. China.

[2] MHURD and AQSIQ, 2006. Technical code for design of outdoor wastewater engineering (GB50014). The Ministry of Housing and Urban-Rural Development (MHURD); the General Administration of Quality Supervision, Inspection and Quarantine (AQSIQ), P. R. China.

[3] MHURD and AQSIQ, 2017. Technical code for urban flooding prevention and control (GB51222). The Ministry of Housing and Urban-Rural Development (MHURD); the General Administration of Quality Supervision, Inspection and Quarantine (AQSIQ), P. R. China.

[4] MHURD, 2018. Examination methods for water quality of municipal sewage (CJ/T51). The Ministry of Housing and Urban-Rural Development (MHURD), P. R. China.

[5] MHURD and AQSIQ, 2014. Technical code for groundwater monitoring (GB/T 51040). The Ministry of Housing and Urban-Rural Development (MHURD); the General Administration of Quality Supervision, Inspection and Quarantine (AQSIQ), P. R. China.

[6] AQSIQ and SAC, 2017. Specifications for surface meteorological observation – Air temperature and humidity (GB/T 35226). The General Administration of Quality Supervision, Inspection and Quarantine (AQSIQ), P. R. China. Standardization Administration of the People's Republic of China (SAC).

A National Standard of the People's Republic of China

Assessment Standard for Sponge City Effects

GB/T 51345-2018

Preparation notice

GB/T 51345-2018 'Assessment standard for Sponge City effects' (this standard), was approved by Ministry of Housing and Urban-Rural Development and published through the public notice on 26 December 2018.

The working group developed this evaluation system for Sponge City effects on the basis of extensive investigation, learning from practical experience, relevant international standards and other advanced foreign standards, plus extensive consultation, taking full consideration on the state-of-the-art of China's Sponge City infrastructure development and the research results achieved from the project 'Water Pollution Control and Remediation' and other key national research projects.

In order to help the broad audience involved in planning, design, construction, research, education and for other institutions concerned to correctly understand and implement the provisions of this standard, the working group has prepared this article description using the same order of chapters, sections and articles. Its intent is to convey the purpose of the article's provisions, theoretical principles, and other matters that need to be paid strict attention to during assessment of Sponge City implementation effects. However, it must be stated that this article description does not have the same legal effect as the standard articles, and therefore is intended only as an aid and a reference to help users understand the key provisions of the standard.

Contents

1 GENERAL PROVISIONS

1.0.1 The development of Sponge City is a crucial measure during the development process of urban transformation in the new era. It can promote the ecological civilization and green development, stimulate supply-oriented structural reform, enhance the transformation of urban development mode, and improve the systematic development of urban infrastructure. Further, learning from advanced international experience during the Sponge City implementation, it aims to establish a set of evaluation systems for Sponge City implementation that is suitable for China's national conditions, to formulate and implement standardized evaluation standards, which is of great significance for guiding the implementation Sponge City development initiative.

1.0.2 This article specifies the application scope of the assessment standard.

1.0.3 The traditional urban development and construction process results in the excessive growth of impervious surface area and over-dependence on traditional grey drainage systems (pipes, culverts, channels, etc.). This, in turn, has destroyed the natural cycle of water systems, altered the hydrological characteristics of catchments and, as a result, has led to substantial impacts on urban water ecology, water environment and water resources. As importantly, it has increased the likelihood of flood events and risk to public safety. The development of the Sponge City, through preservation of the natural ecological systems of mountains, rivers, lakes, forests, farmland and grassland, can strengthen the management of rainfall runoff, retain maximally the hydrological characteristics before and after urban development, restore and rehabilitate water ecology, protect water environment, and preserve water resources and enhance the capability of urban disaster prevention and mitigation.

1.0.4 The traditional urban development method relies mainly on the traditional grey pipe drainage networks, leading to the loss of the natural hydrological functions, such as infiltration, retention and purification of rainfall runoff, namely, the loss of the natural 'sponge' function of urban catchment.

The technical concept of Sponge City development has changed from traditional 'end-of-pipe remediation' to 'source reduction, process control and systematic remediation'. The control and management method has changed from traditional 'quick discharge' to a holistic approach of 'infiltration, detention, retention, purification, utilization and discharge'. By restoring the 'sponge' function of the urban surface, it not only relieves the pressure of ecology, environment and resources, but also reduces the cost of construction, operation and maintenance through the combination of grey and green infrastructure measures.

2 TERMS AND SYMBOLS
2.1 Terms

2.1.1 Sponge City is a concept for systematic management of urban water-related problems. The key target is to achieve this new urban stormwater management approach through the process of planning, design, construction, operation and maintenance to control effectively urban rainfall runoff. Through implementation of combined measures of infiltration, detention, retention, purification, utilization and discharge, it aims to provide a foundation to achieve the multiple objectives of controlling runoff volume, reducing flowrates and pollutants, thereby alleviating urban flooding, controlling runoff pollution, improving the water environment and water ecology. All of this can be characterized as important backbone for achieving comprehensive management of mountains, rivers, forests, lakes and swales, for green development which leads over time to the building of a beautiful China.

2.1.2 Through use of rainfall infiltration, retention and purification of natural catchment surface, rainfall runoff can be controlled, namely, a natural surface can act like a sponge to absorb, infiltrate, retain and purify the runoff for different types of utilization, before it is discharged. The hydrological features can also be achieved through artificially constructed catchment surfaces and facilities, which is an important artificial sponge object.

The hydrological cycle includes the processes of precipitation, evapotranspiration and runoff. At urban scale, the hydrological cycle is mainly reflected by rainfall and runoff. The traditional urban development and construction mode has led to excessive growth of impervious surface areas, altered the natural hydrological characteristics and modified the natural water circulation route, resulting in considerable impacts on urban water ecology, water environment and water resources, and also increased the risk of water disasters. Since the impact of urban development and construction on the hydrological cycle is mainly on runoff, the purpose of Sponge City infrastructure development is to protect and restore natural hydrological characteristics in the urban areas. The core focus is on restoration of the natural rainfall runoff process and, thereby, control of urban runoff.

Under natural conditions, significant surface runoff is usually generated during heavy rainfall, namely those storm events with small probabilities of occurrence; while during small and medium rainfall events, i.e. the ones with higher probabilities of occurrence, a large amount of surface runoff is less likely to occur. Runoff is mitigated by infiltration and retention on the natural catchments' surface. Thus, to control frequent runoff, small and medium rainfall events with high probability of occurrence, should be controlled first. The frequency of occurrence of small- and medium-sized rainfall events is high, and the cumulative rainfall from these events accounts for a large proportion of the total annual rainfall, leading to high probability of runoff events and major pollution load contribution throughout a year. The volume capture ratio of annual rainfall is the ratio of the captured and the controlled annual rainfall to the total annual rainfall. This reflects the degree of controlling rainfall runoff by natural and artificial sponge facilities under the unsaturated conditions, and it can also reflect the level of capturing and controlling runoff from a large number of small and medium rainfall events. Therefore, it is of great significance to preserve the hydrological characteristics as the fundamental status of ecology and to achieve the comprehensive goals of Sponge City development.

Based on the multi-year rainfall data, the corresponding relationship of the volume capture ratio of annual rainfall and the total rainfall can be established, and the design rainfall depth can be determined as a key parameter for the design of runoff volume control facilities.

Based on 24-hour precipitation data for multiple years (no less than 30 years), and by excluding the rainfall events which are less than or equal to 2 mm and all snowfall data, each of the 24-hour rainfall events is used to draw a curve of rainfall depth vs. rainfall events (Figure 1). The X-axis is the cumulative number of rainfall events (sorted by 24-hour rainfall depths from small to large) for more than a 30-year period, and the Y-axis is the rainfall depth of the corresponding rainfall events.

According to the curve in Figure 1, the design rainfall depth H corresponding to the volume capture ratio of annual rainfall α, which can be calculated by the formula: $\alpha = (C_1 + C_2)/(C_1 + C_2 + C_3)$.

According to the curve in Figure 1, a curve relating the capture ratio of annual rainfall α and the design rainfall depth H can be derived (Figure 2).

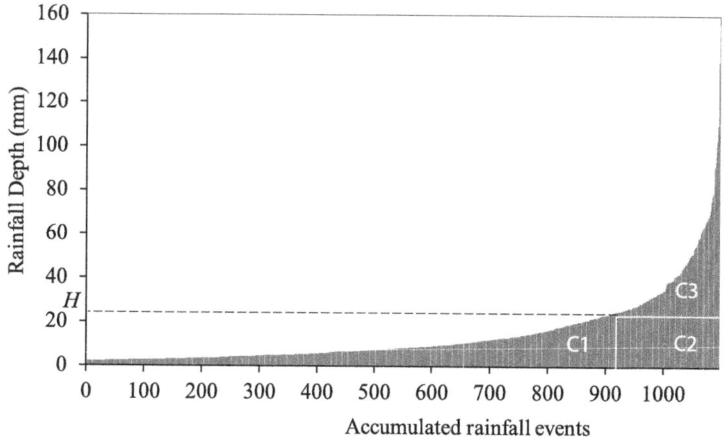

Figure 1 Schematic diagram of the relationship between rainfall events and rainfall depth.

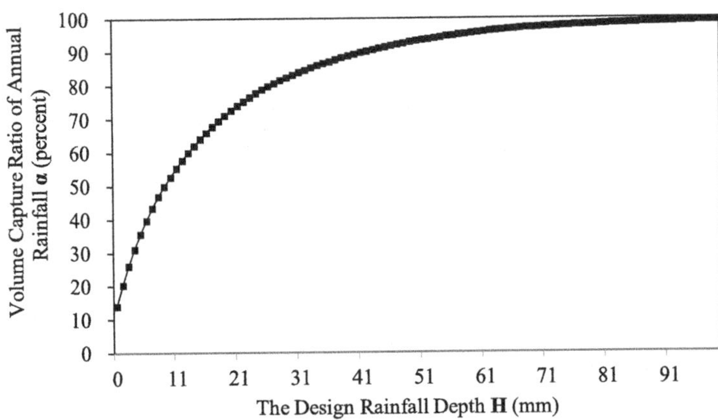

Figure 2 Schematic diagram of the relationship between the volume capture ratio of annual rainfall α and the design rainfall depth H.

According to the design rainfall depth corresponding to the volume capture ratio of annual rainfall and catchment area controlled by the Sponge City infrastructure, the runoff control volume can then be calculated as the design control volume of the runoff control facility.

2.1.3 During urban development, the sponge function in controlling of runoff volume, water quality, runoff peak value and runoff frequency should be fully utilized, to protect and maximize rehabilitation of the hydrological characteristics of the natural catchment surface, and to achieve the multiple objectives in controlling of surface ponding and local flooding, as well as pollution control.

2.1.4 A catchment is an area delineated according to the topography, terrain or the layout of the drainage network, overland and underground sewers runoff routing, also including, ditches and channels, detention and retention facilities and urban water bodies. A catchment area can be further divided into sub-catchments.

The area of collecting overland rainfall runoff that is determined by topography or drainage boundary can also be called 'watershed' or 'basin'.

2.1.5 Traditionally, runoff is drained from the urban areas by sewer systems, which does not utilize the 'sponge effect' of urban surface. However, under the Sponge City development, the rainfall runoff will be treated through the measures of infiltration, detention, retention and purification, such that the excess runoff will be discharged naturally only after the sponge body is saturated.

The overflow outlets in this standard refer to the facilities that drain runoff that exceeds the control capacity of the sponge body to downstream recipients. According to the definition of Basic Term Standard for Water Supply and

Drainage Engineering (GB/T50125), a discharge outlet is a physical facility that drains the runoff, or treated wastewater, to receiving water bodies.

2.1.6 Green infrastructure defined in this standard refers to green stormwater infrastructure facilities, including sunken green beds, bioretention, stormwater detention or retention ponds, etc.

2.1.7 Grey infrastructure includes reinforced concrete drainage pipes and channels, pumping stations and other drainage engineering facilities with high energy consumption for production or operation.

2.1.9 The urban waterbody defined in this standard includes but Does not limit to the various receiving waterbodies of urban drainage systems within the urban development boundary, but it does not include waterbodies within residential areas.

3 BASIC REQUIREMENTS

3.0.1 This article specifies the objects and items to address when assessing effectiveness of Sponge City implementation.

'Urban built-up area' refers to the area where the urban administrative area has been well developed and where municipal and public service facilities are in place.

Sponge City implementation should start in areas experiencing severe problems such as water ponding, local flooding and water pollution. Using findings of water ponding, stormwater discharge outlets and CSO problems, investigations of them need to proceed upstream, correctly delineate catchments and sub-catchments, identify Sponge City implementation options and then proceed to make plans for implementing projects to achieve integrated regional or river basin management, with the goals of attaining 'no surface ponding during small rain events, no local flooding during heavy storm events, no malodorous waterbodies and alleviate heat island effects'. This will require evaluation of the sponge effects of the source reduction facilities on entire drainage catchments and their built-up areas.

3.0.2 The article specifies the evaluation results of Sponge City implementation. The urban build-up area shall be validated though the use of the data in the 'China Urban Statistics Yearbook' published in the year of evaluation.

3.0.3 This article specifies the assessment criteria of Sponge City implementation effects. The requirements for the examination and the inspection are described in item 4.0.2. Sponge City implementation plays an important role in alleviating the decline of groundwater levels and of urban heat island effect. However, since both indicators are affected by multiple factors, which causes uncertainties to the assessment. Thus, it is difficult to determine accurately the trend of changes based on short-term monitoring, and further, it is difficult to establish quantitative or qualitative correspondence of Sponge City implementation with other related factors. Therefore, although the Variation Trend of Groundwater Depth and Urban Heat Island Effect are to be assessed, but the conclusion of the assessment should not affect the judgment of the evaluation's results.

3.0.4 This article specifies the overall assessment methods and conditions. Hydrological characteristics are represented by the three typical characteristic years: the wet year, the medium year and the dry year. However, a single hydrological cycle period is set as one year. As a result, the standard requires continuous monitoring for at least one year. Longer period monitoring is encouraged whenever possible.

Urban stormwater projects are typically designed based on the statistics of urban hydrology. Due to the high variability of runoff quantity and quality data, it is not realistic to use a large number of actual storm events to evaluate the designed conditions of engineering facilities. As a result, it is suggested that assessing the effects of the Sponge City should be done using a comprehensive evaluation approach that includes monitoring, model simulation, reviewing design and construction documents and on-site inspection.

3.0.5 This article specifies the selection of typical projects and the requirements for monitoring to assess implementation effectiveness of all source reduction installations. In order to save cost and time of evaluation and improve evaluation efficiency, it is essential to select representative projects for monitoring and evaluation, which represent the experience of local Sponge City implementation practices of the representative projects. This will also provide data for calibration and validation of hydrological and hydraulic assessment models that are used to help with the overall urban water environment and flood control assessment. Further, when it is possible, typical facilities should be selected for monitoring within the monitoring project, which should have clear catchment boundaries and for which monitoring equipment can easily be installed and can provide the data that support model parameters and inputs being used.

Representative projects mainly include residential areas, roads, parking lots, open squares, public parks and protective green spaces. Among them, when a residential area is selected for monitoring, it includes the projects in commercial and industrial areas.

The selection of representative monitoring projects should consider the following: (1) It is within the same catchment. (2) It has significant potential effectiveness to solve the problems of surface ponding, runoff pollution and pollution caused by CSOs. (3) The technical measures adopted in the project and the project scale are representative. (4) The pipe network information is complete, and pipe defects have been detected and repaired.

4 ASSESSMENT ITEMS

4.0.1 This article specifies the specific assessment items and requirements of Sponge City development. The goal of Sponge City development is to be achieved through the restoration of the natural hydrological characteristics, which are evaluated mainly for the four following stormwater aspects: runoff volume, peak flow, frequency of runoff and water quality, which are also the main issues when assessing Sponge City implementation effects.

(1) Volume capture ratio of annual rainfall and runoff volume control

The urban new development area refers to areas dominated by new development projects. It is relatively easy to implement the volume control requirements during the urban planning and design stages of new developments. The runoff volume control target should be decided based on the principle of achieving hydrological characteristics under natural ecological conditions. The volume capture ratio shall not be smaller than the lower limit specified according to the 'Zoning map' (Figure 4.0.1).

The urban re-development area refers to areas dominated by reconstruction and extension projects. For existing urban built-up areas, the implementation of runoff volume control target will vary from project to project. Its size should be based on what is needed to solve problems such as urban surface ponding, local flooding, runoff pollution and pollution by CSOs. The scale of this volume should be based on technical and economic assessments as well as the conditions of the areas being renovated or expanded. Wherever possible and applicable, the built-up areas should adopt the runoff volume control targets set out for new development areas, pursuing the preservation of hydrological characteristics to the maximum degree, ones that would exist under natural ecological conditions.

The volume capture ratio of annual rainfall can be determined according to the rainfall runoff coefficient in the natural state of the region. It is calculated according to the equation (5.1.3) of this standard. If the local hydrological information is incomplete, the volume capture ratio of annual rainfall can be determined according to Figure 4.0.1 of this standard.

In dry areas, natural infiltration capacity is large relative to the annual rainfall depths, the volume capture ratio of annual rainfall may reach the upper limit; while in wet areas, the groundwater level may be high and the permeability can be limited. Therefore, the lower limit may apply.

(2) Implementation effectiveness of a source reduction project

The effectiveness of implementing a source reduction project is very important for achieving overall development effects in urban build-up area. Therefore, this standard requires a comprehensive evaluation of how effectiveness of the implementation for residential areas, roads, parking lots, open squares, public parks and protective green space.

(i)   For residential areas. The drainage network should be designed to best address the topography and geomorphology. Stormwater runoff should be collected on land surfaces as much as possible, and the use of pipe networks should be reduced or, where possible, disconnected from surface stormwater runoff. Runoff control should be achieved through 'infiltration, detention, retention, purification and utilization'. When the control requirements for runoff volume, peak flowrate and pollution control have been met, the system design should be designed to permit all overflows to be discharged into the drainage networks.

In practice, for some of the new or redevelopment projects, due to insufficient space and limitation of vertical conditions, difficult construction and high construction cost, it may be difficult to meet the full requirements of the low limits specified according to the 'Zoning map of volume capture ratio of annual rainfall in China (Figure 4.0.1 of this Standard)'. In such conditions, it will be necessary to perform technical and economic analysis for the specific project conditions, and then decide the value of volume control ratio. Based on specific site conditions, this standard allows the projects to meet similar control requirements of urban planning, while also meeting the assessment requirements of this standard, where relevant urban planning refers to the special urban planning for Sponge City implementation and to other detailed urban control planning.

Many national and international studies and practices have shown that a significant part of runoff pollution is generated by small and medium rainfall events. Because of the random nature and complexity of runoff pollution variations, runoff pollution is generally managed through the control of runoff volume and the use of drainage infrastructure to reduce pollutants from that volume.

Rainfall runoff pollution is driven by atmospheric dusts, vehicle exhaust, catchment characteristics and many other factors affecting the urban landscape. The composition of pollutants is very complex. Suspended Solids (SS) often have certain correlation with other pollutants, therefore the Suspended Solids (SS) parameter is used an indicator for runoff pollution levels. Each city should establish the correlations of SS and other pollutants based on monitoring and analysis of typical catchment or under various land-use conditions.

Annual reduction rates of SS in runoff are related to the initial values of SS concentration, the first flush efficiency of the runoff, and the facility's ability to remove SS. The concentration of SS of rainfall runoff is generally high in China, and the first flush of runoff at the source is often more efficient than control at end of the pipe collection systems. As a result, the SS in the initial runoff can be efficiently removed by the source reduction facilities, where reduction rates can be relatively high. In the 'Technical Guideline for Sponge City implementation – Low Impact Development Stormwater System Construction', the annual SS removal rate was estimated according to the product of volume control ratio of annual rainfall and the facility SS reduction rate. However, this calculation method does not take into account the effect of first flush of runoff, resulting in smaller removal rate of annual SS than may actually occur. It is therefore suggested that cities through monitoring establish the relation curves of SS concentration for different types of urban land uses or surface conditions under runoff flowrates produced by different rainfall events. This monitoring will permit estimating SS Event Mean Concentration (EMC) under different rainfall conditions, estimates of the annual SS reduction rates related to the runoff volume of control facility's SS removal rates and also related to the runoff control volume ratio.

The annual TSS reduction rates by stormwater quality control facility in US are typically between 80 percent and 95 percent. Considering the actual condition of runoff pollution in China, the standard needed to ensure adequate SS capacity, specifies that the volume control rate of annual runoff for new developments be no less than the lower limit defined according to the 'Zoning map of the volume capture ratio of annual rainfall in China' (Figure 4.0.1), and that the SS removal rate be no less than 70 percent. For redevelopments, the SS removal rate should be not less than 40 percent. The redevelopment projects should increase over the exiting conditions the control ratio of impervious surface area and the corresponding volume control ratio of rainfall runoff.

In practice, when it is difficult to control runoff pollution through runoff volume, physical treatment methods such as sand removal and filtration may also be used. Further, to ensure the overall runoff pollution control objective of the project, measures should be taken to minimize surface runoff of all impervious surface areas, especially roads, parking lots and surfaces that have high potential risk of runoff pollution.

In addition to climatic factors, the main impact variables of hydrological characteristics prior to the new development, include impervious surface area, topography, soil characteristics, etc. When data for these variables are lacking or impossible to be used to represent the benchmark condition for hydrological analysis before the development, a ratio of 5% can be used as the baseline ratio of the impervious surface area to the total area of the project site. The topography and soil characteristics can also be reasonably \ assumed based on relevant data or post-development conditions.

Generally, the green space percentage for the second level residential areas is 30 percent – 35 percent, the building density (percentage of the roof area to the total area) is 35 percent – 40 percent, and the impervious land surface area proportion, excluding buildings, is about 25 percent – 35 percent. And, the effective impervious ratio, after excluding roof areas, of the total surface area is about 42 percent – 54 percent. Thus, it is encouraged to convert some impervious land surface areas to permeable pavements. The standard recommends that the impervious ratio for the new development community shall not exceed 40 percent.

(ii) Roads, parking lots and open squares are areas that are dominated by impervious surfaces (such as pavement) and are the major areas that contribute to rapid generation of runoff that lead to surface water ponding, local flooding and runoff pollution. In order to reduce the impacts of the runoff quantity and pollution problems on urban ecology and environment, it is important to control runoff volume, peak flowrate and runoff pollution through Sponge City implementation measures. For new development projects, measures such as physical and ecological treatment processes should be applied to control rainfall runoff on roads, parking lots and open squares. For redevelopment projects, use of control standards for rainfall runoff by the requirements for new projects is recommended.

(iii) Parks and protective green spaces: when planning and designing new and renewal of parks and green space projects, it shall not impair or reduce the main functions of green space as entertainment parks and emergency shelters, etc. Through receiving water from surrounding areas, they can contribute to solving regional water problems, such as water ponding, flooding, runoff pollution and pollution caused by Combined Sewer Overflows (CSOs) by playing a role in runoff control, flood detention or retention. In practice, because of the difference in dimension, vertical conditions and main functions of parks and protective green spaces, it is difficult to require all of them to receive rainfall runoff from surrounding areas. Therefore, this standard proposes that parks and protective green spaces should have capacity to receive rainfall runoff from surrounding areas according to the requirements of planning and design before being used for runoff management purposes.

(3) Control of road water ponding and local flooding

By reduction of urban runoff at the source, the peak discharge of runoff can also be reduced and delayed and the presence of local flooding can be reduced. At the same time, by making use of the capture capacity of mountains, rivers, lakes, forests, farmland and grassland and vertical control and control system for extreme storm events, the risk of local flooding can be reduced.

Through Sponge City implementation and implementation of grey and green solutions, the urban sewer system (minor drainage system) and flood control system (major drainage system) can meet the requirements of the 'Technical code for design of outdoor wastewater drainage system' (GB 50014)[2] and the 'Technical code for urban flooding prevention and control' (GB51222)[3].

(4) Urban water quality

Runoff pollution during rainfall and pollution by overflow discharges from separate sewer systems because of hydraulic overloads, infiltration and inflow or mis-connections and from CSOs are one of the major pollution sources to urban water bodies. Controlling of rainfall runoff through Sponge City implementation can not only reduce the problems of runoff pollution, misconnection of the separate sewer systems and CSOs, but also help to solve the problems of mis-connection of the separate systems and reduce CSOs from sources of rainfall runoff.

Source reduction, pollution treatment, ecosystem restoration, living water and long-term remediation is a systematic approach to address the problems of malodorous waterbodies. Sponge City implementation can play a significant role in controlling of runoff pollution, pollution by CSOs, restoration of ecological shoreline and downstream water purification and ensuring safe water quality. Moreover, implementation of combined grey and green infrastructure solution is intended to reduce the costs of construction, operation and maintenance.

The annual overflow volume control ratio during rainfall refers to the ratio of the controlled overflow volume by a series of mitigation measures, which includes correction of mis-connection, interception, detention and retention and treatment to the total overflow volume. Among them, detention and retention facilities include bioretention, rainwater detention or retention ponds, basins and tanks. The treatment facilities refer to the wastewater treatment plants and the overflow treatment stations. The treatment process includes 'primary treatment + disinfection', 'primary treatment + filtration + disinfection' and 'detention + constructed wetlands' and the whole processes of wastewater treatment plants.

Recognizing the high variability of rainfall characteristics, the operation of sewer systems, the water environmental capacity of the receiving water bodies and the background information of overflow pollution of different regions in China, this standard suggests overflow pollution control criteria that should be decided based on the technical and economic analysis. The control indicators, in addition the overflow volume control ratio, may include annual average overflow frequency and the reduction rate of total pollution load.

Due to the limited experience of the overflow pollution control and lack of data, this standard refers to the experience and practices of the CSO controls in the United States, in combination with China's national condition, specifics of addressing through pollution control indicators and criteria mis-connections of separate sewers systems and of CSOs.

The control standard of overflow regulation in United States is based mainly on controlling of the average number of annual overflows or on annual overflow volume reduction ratio, the annual reductions in total suspended solids (TSS) or biochemical oxygen demand (BOD), fecal coliform, pH, SS, BOD, and/or DO concentrations. The average annual overflow frequency of CSOs in several U.S. cities is set at one to four times, the annual overflow volume control rate ranges from 80 percent to 90 percent. As an example, the average monthly limits for TSS effluent concentration for Cities of Philadelphia and Portland of the USA are set to 25 mg/L and 30 mg/L respectively. In a pilot city of Sponge City implementation in Southern China, the control standard for average annual overflow frequency is set to not exceed 15 times, and the annual overflow volume control ratio is about 70 percent.

(5) Natural ecological pattern management and water ecological shoreline

According to the 'Specification of Urban Water System Planning' (GB 50513), ecological shoreline refers to the natural shoreline preserved for the protection of the ecological environment or the shoreline with natural features after ecological restoration.

The water ecological restoration includes complex aspects, such as restoration of ecological baseflow, restoration of biodiversity and habitat creation. Being a preserved space for interception of pollution and water purification at the end of urban drainage system, ecological shoreline is an important aspect of water ecological restoration. Therefore, this standard proposes the evaluation requirements for water ecological shoreline protection.

(6) Variation trend of groundwater depth

The impervious pavement in the city cuts off the natural path for stormwater infiltration, reduces stormwater infiltration and groundwater recharge, resulting in the drop of groundwater levels. The implementation of Sponge City concept could result in stormwater runoff being recharged into the ground or discharged, after treatment, to the river in order to maintain the river's baseflow.

(7) Urban heat island effect

Sponge City implementation through implementation of runoff control measures, such as increasing of permeable surface pavement and natural vegetation, restoration of the natural hydrological cycle, thus plays an important role in alleviating the urban heat island effect.

4.0.2 This article specifies the examination and inspection items. The sponge city implementation can increase the permeable surface areas of the city, replenish groundwater, and alleviate effectively the problem of groundwater level drop. However, the variation of groundwater level is affected by many other factors, such as hydrogeological condition, climate change and human activities, etc. Therefore, it is essential to make comprehensive analysis of the changing mechanics based on long-term operation of groundwater level and the change of the water environment.

The primary factors leading to the urban heat island effect include the increase of the impervious surface and the decrease of natural vegetation coverage, the heat emission from human activities, such as vehicle waste gas emission, and the impact of regional climate change. Sponge City implementation aims to guide better protection of natural vegetation, and increase the permeable surface during urban development, which can alleviate effectively the urban heat island effect, though it may be affected by other factors.

5 ASSESSMENT METHOD
5.1 Volume capture ratio of annual rainfall and runoff volume control

5.1.2 For the Sponge City infrastructure that either filters water before discharge or cause it to be infiltrated in the ground, the effective retention volume V_s refers to the volume above the infiltration and filtration surface; V_s for extended retention or detention facilities refers to the volume at bottom of the larger basin that is specifically sized for runoff pollution control; the V_s for the retention tank and water body refers to their storage volume while excluding the volume specifically for peak flow controls.

When SS is removed mainly through the process of sedimentation, the design emptying time of the extended retention or detention facilities is determined according to the sedimentation time required to ensure sufficient removal of SS. If the data is not available, a default value of 40 hours may be used. In the areas with high heavy metal pollution levels, such as gas stations and urban roads, a default value of 72 hours may be taken. When SS is removed mainly through the process of filtration, the design emptying time of the extended retention or detention facilities is determined according to the design emptying time of bioretention systems or sand filtration systems.

The hydraulic conductivity should be determined by the soil or artificial media whose hydraulic conductivity determines the infiltration and retention ability of the facilities, after accounting for surface clogging over time.

5.1.3 For the pervious land surface that is not controlled by any facility, such as porous pavements and green spaces, the 'Zoning map' (Figure 4.0.1 of this standard) can be drawn by referring to the provision 2.1.2 of the article description in this standard. The runoff coefficient refers to the ratio of the annual average runoff volume to the volume of annual average rainfall, which is the 'annual runoff coefficient'. If the data is not available, it should be determined according to runoff coefficients for different surface types specified in the current national standard 'Technical code for rainwater management and utilization of building and sub-district' (GB50400)[1].

5.1.4 The monitoring project is determined according to the provision 3.0.5 of this standard. In the cases of extreme dry or wet years, the monitoring duration may need to be extended for one additional year according to the weather condition.

5.1.5 The pipe defects refer to the structural and functional defects, the inspection of pipe defects may follow the provisions of the current industrial standard 'Technical specification for operation, maintenance and safety of sewers & channels and pumping stations in city and town' (CJJ 68).

5.2 Implementation effectiveness of the source reduction project

5.2.1 Runoff pollution varies with the rainfall events, which have high randomness and complexity, thus the control of runoff pollution is generally achieved through control of runoff volume. The technical requirements and costs are high when extensive monitoring is required to assess the runoff pollution control effect. Based on the volume capture ratio of annual rainfall and the runoff volume control, the assessment of runoff pollution control is done through assessing the facility's pollution control capacity as well as the ratio of impervious surface that is controlled by facilities. The factors that impact the pollution control capacity of a facility include design configuration, runoff control volume, emptying time, operational conditions and selection of vegetation types.

The emptying time is one of the important factors that affect the pollution removal ability of a facility. It is determined by considering different aspects, including effect of TSS removal, runoff volume control, wet tolerance of plants, mosquito breeding issues, etc. In practice, the design emptying time for bioretention facility is generally taken as 12-hours. Considering the surface clogging during the operation of the facility, the saturated hydraulic conductivity of the surface planting soil should result in a conservative design to ensure that the actual emptying time at the initial stage of the facility operation is not more than 6-hours, thus the type of soil and artificial media compositions shall be designed to do so. The design emptying time for extended retention facility may be taken as 40 hours, 70 hours, or the design emptying time for bioretention and sand filtration facilities. When the actual emptying time of the facility is found to be longer than designed value, need for maintenance is clearly indicated.

The first flush effect at source of the impervious surface is usually significant, indicating that high efficiency of pollution control may be achieved if the initial runoff volume is controlled. Studies and practices in China and other counties have shown that the source reduction facilities designed according to the design rainfall corresponding to the volume capture ratio of annual rainfall can collect about 80 percent of SS in the initial rainfall runoff. Accordingly, if SS removal rate of a facility achieves 85 percent~90 percent for the new development project, at least 70 percent of the annual SS loads generated on all the impervious surface may be controlled, provided that the volume capture ratio of annual rainfall achieves the lower limit according to the 'Zoning map' (Figure 4.0.1 of this standard); For redevelopment projects, at least 40 percent of annual SS load reduction may be achieved if more than 60 percent of the impervious area is controlled and the volume capture capacity is maximally achieved within the project.

5.3 Control of road surface water ponding and local flooding

5.3.2 Refer to 'Notice of the Ministry of Housing and Urban-Rural Development of PRC on List of Persons responsible for national urban drainage flood control and key flood-prone sites in 2018' (Urban Construction Letter [2018] No. 40) for the key flood-prone sites. In addition, each city shall continuously update information about new flood-prone sites based on a combined approach of on-site inspection and model simulation.

Drainage design manuals of many cities in the United States have design requirements for the depth of road surface ponding and the submerging width of the road. For example, in city of Denver, the ponding depth of urban main roads should not exceed 15 cm under the 'minor' design storm event for urban drainage system to ensure that for a two-way road there is at least one lane without water ponding, and the maximum allowable width of water ponding does not exceed 2 lanes. By referring to the design requirements of the cities like Denver in US, this standard proposes that in the discharge design of road ditches and the low points of the key flood-prone sites, the hydraulic head or runoff depth used for hydraulic calculations should not be higher than 15 cm during 'minor' storm, and the assessment of actual flooding condition of the key flood-prone points should be done based on this.

5.3.3 The monitoring project is determined according to the provision 3.0.5 of this standard.

5.4 Urban water quality

5.4.4 The setup of the monitoring cross section, monitoring point, sample point for determining urban water quality should follow the relevant provisions of the current industrial standards 'Technical Specifications Requirements for Monitoring of Surface Water and Waste Water' (HJ/T 91) and 'Water quality Technical regulation on the design of sampling program' (HJ 495). The classification of precipitation is determined according to the current national standard 'Grade of precipitation' (GB/T 28592).

5.5 Natural ecological pattern management and shoreline for ecology conservation

5.5.2 The ecological shoreline rate for new and redevelopment projects is defined as ratio of the length of the ecological shoreline to the total length of shoreline, excluding the length of shoreline reserved for production and flood protection.

IWA Publishing's authorised EU representative for General Product
Safety Regulations is Diane D'Arras, 15 rue Duret, 75116 Paris,
France, e-mail: safety@iwap.co.uk.

Printed and bound by CPI Group (UK) Ltd, Croydon, CR0 4YY

16/06/2026

02140390-0001